found

BEACHCOMBER

by George Mackay Brown

Monday I found a boot –
Rust and salt leather.
I gave it back to the sea, to dance in.

Tuesday a spar of timber worth thirty bob.
Next winter
It will be a chair, a coffin, a bed.

Wednesday, a half can of Swedish spirits.
I tilted my head.
The shore was cold with mermaids and angels.

Thursday I got nothing, seaweed
A whale bone,
Wet feet and a loud cough.

Friday I held a seaman's skull,
Sand spilling from it
The way time is told on kirkyard stones.

Saturday a barrel of sodden oranges.
A Spanish ship
Was wrecked last month at The Kame.

Sunday, for fear of the elders,
I sat on my bum.
What's Heaven? A sea chest with a thousand gold coins.

From 'Fishermen with Ploughs' 1971, John Murray,
and 'The Collected Poems of George Mackay Brown' 2005, John Murray.
By kind permission of the Estate of George Mackay Brown.

found

BEACHCOMBING IN ORKNEY

KEITH ALLARDYCE

Published by The Orcadian Limited (Kirkwall Press)

Hell's Half Acre, Hatston, Kirkwall, Orkney, KW15 1DW

Tel. 01856 879000 • Fax 01856 879001 • www.orcadian.co.uk

Text: Keith Allardyce © 2012

Photographs: Keith Allardyce © 2012

ISBN 978-1-902957-51-7

Printed by The Orcadian Ltd, Hatston Print Centre,

Hell's Half Acre, Kirkwall, Orkney, Scotland, KW15 1DW

Proceeds of the Royalties from *Found* will be donated to Stromness Museum

ACKNOWLEDGEMENTS

Without the kindness and good humour of so many people during my journey around the Orkney Islands, this book would not have been made.

Therefore, I wish to thank, firstly, all the people who appear in this collection of portraits, who showed me their sea-finds, and who shared their stories so readily with me.

For his support and advice throughout the *Found* project, and for writing the introduction to this book, I am indebted to Bryce Wilson, writer, historian, artist, and formerly Orkney Museums Officer. On reading his first draft, I wrote to Bryce to say how much I appreciated it - commenting on the flow and the alliteration. He replied with his usual modesty '....it was my happy task to link the gems of others long gone, dredged and cast upon the shores of poetry and prose. (There I am, off again)'.

And my thanks, too go to Tom Muir, writer and much-travelled storyteller, and Exhibitions Officer at The Orkney Museum, for his support and advice throughout the project, and for his Foreword. Having been invited to contribute the essay, he said he was grateful for 'making me think back to how the sea shaped me during my childhood, and the long forgotten memories that it brought back'.

My debt of gratitude also goes to Doris Stout, doyen of Stromness and Deputy Lord Lieutenant of the county, who continues to contribute so much to the Orkney community, and who made time to contribute to the *Found* project.

There was much hospitality and humour served in equal measure at The Haven when Doris and her now late husband John lived at the Northern Lighthouse Board's north of Scotland H.Q. in the centre of Stromness; and this great hospitality and humour continues at Doris's new home on the outskirts of the town.

So my thanks go to Doris for inviting me and my partner Ikuko to stay with her during the *Found* project; and for her great tolerance, during our stay, when I returned to Norhaven so often with unsavoury finds from the Orkney shore (I see all those dolphin bones and bird skulls as pieces of modern art, sculpture...).

Finally, I would like to thank my partner, Ikuko Tsuchiya, Associate Lecturer in Fine Art at Teesside University, for travelling around Orkney with me, and for all her support in many ways throughout the project. 'Orkney has been for me the perfect balance of people and landscape', she told me, after our first visit together.

Keith Allardyce.

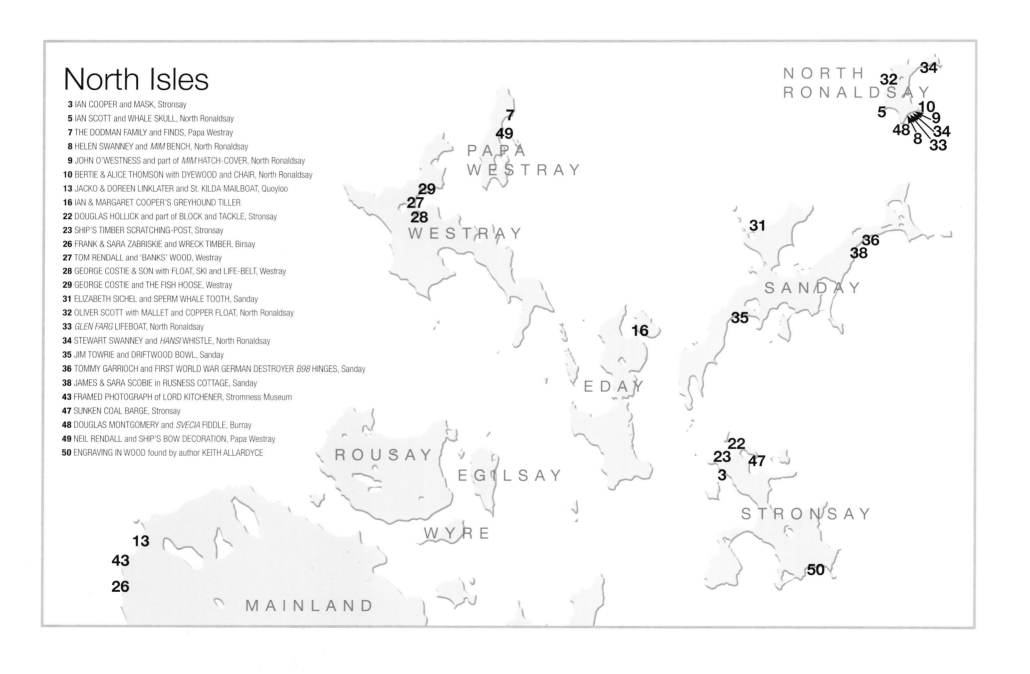

North Isles

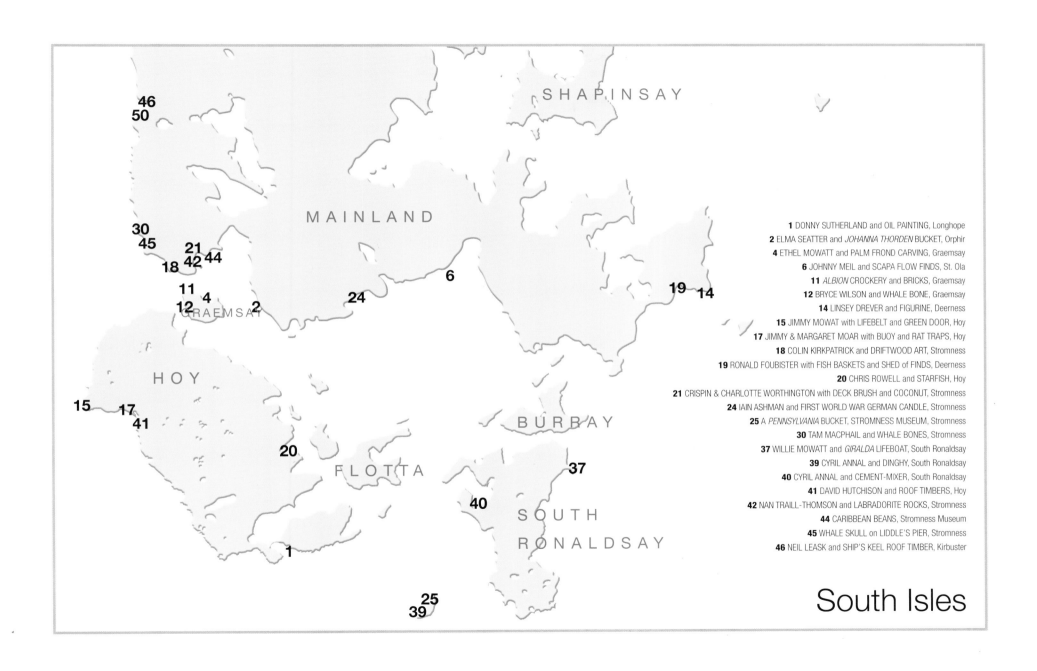

SHAPINSAY

MAINLAND

46
50

30
45

21
42 44

18

11
4

12 GRAEMSAY

2

6

24

19 14

HOY

15

17
41

20

BURRAY

FLOTTA

37

40

SOUTH

RONALDSAY

1

25
39

1 DONNY SUTHERLAND and OIL PAINTING, Longhope
2 ELMA SEATTER and *JOHANNA THORDEN* BUCKET, Orphir
4 ETHEL MOWATT and PALM FROND CARVING, Graemsay
6 JOHNNY MEIL and SCAPA FLOW FINDS, St. Ola
11 *ALBION* CROCKERY and BRICKS, Graemsay
12 BRYCE WILSON and WHALE BONE, Graemsay
14 LINSEY DREVER and FIGURINE, Deerness
15 JIMMY MOWAT with LIFEBELT and GREEN DOOR, Hoy
17 JIMMY & MARGARET MOAR with BUOY and RAT TRAPS, Hoy
18 COLIN KIRKPATRICK and DRIFTWOOD ART, Stromness
19 RONALD FOUBISTER with FISH BASKETS and SHED of FINDS, Deerness
20 CHRIS ROWELL and STARFISH, Hoy
21 CRISPIN & CHARLOTTE WORTHINGTON with DECK BRUSH and COCONUT, Stromness
24 IAIN ASHMAN and FIRST WORLD WAR GERMAN CANDLE, Stromness
25 A *PENNSYLVANIA* BUCKET, STROMNESS MUSEUM, Stromness
30 TAM MACPHAIL and WHALE BONES, Stromness
37 WILLIE MOWATT and *GIRALDA* LIFEBOAT, South Ronaldsay
39 CYRIL ANNAL and DINGHY, South Ronaldsay
40 CYRIL ANNAL and CEMENT-MIXER, South Ronaldsay
41 DAVID HUTCHISON and ROOF TIMBERS, Hoy
42 NAN TRAILL-THOMSON and LABRADORITE ROCKS, Stromness
44 CARIBBEAN BEANS, Stromness Museum
45 WHALE SKULL on LIDDLE'S PIER, Stromness
46 NEIL LEASK and SHIP'S KEEL ROOF TIMBER, Kirbuster

South Isles

INTRODUCTION

Through the heaped mysteries of waith and wrack,

When the long wave from the long beach draws back,

I wander where the Ocean weaves her spells

With seaweed and with shells.

Here, 'mid the changes of the changeful tides,

Save the brown rocks naught tarries, naught abides;

Changeful as Fortune, here the restless sea

Casts up her spoils for me. [1]

Beachcombing is an honorable tradition. Duncan Robertson's poem eloquently expresses the joy of wandering the shores, on the lookout for the useful, the beautiful and the mysterious. (There is honour among beachcombers. Finds laid on the banks above the high water mark are the property of the finder, and may not be removed by any other person.)

Orkney's earliest settlers were beachcombers. They supplemented their diet with cockles and limpets, oysters and crabs and seaweeds. Limpets and other shells made scoops and spoons and simple bow drills. Sometimes the ocean would offer up a great tree, swept by floods and storms from distant lands. (Log boats have been found in island lochans.)

The grounding of a great whale was a time of rejoicing. Here was flesh for a feast, oil aplenty for the lamps, bone to make tools and implements. Huge vertebrae could serve as seats by the fire, or be hollowed to make useful containers.

Late in the 17th century, the Rev James Wallace wrote of 'many spout Whales or Pellacks which sometime run in great numbers upon the shore and are taken ... In many places they get Cockles in such abundance, that of the shells a great dale of Lime is made; excellent for Plaistering'. [2]

Currents warmed in southern seas bore mysterious objects. Wallace recorded the finding of "Molluca beans", smooth shiny seeds of tropical shrubs growing in Caribbean islands. '... these pretty Nutts, of which they use to make Snuff Boxes, there are four sorts of them ... sometime ... they find live Tortoises on the shore ... also at some times is to be found Spermaceti, Amebergreise [wax from the head of the sperm whale, highly valued in cosmetics and leatherwork, and as a lubricant] ... Sometimes they find exotick fowls driven in by the wind in time of a Storm: I myself saw one that had a long Beak, a large tuft on the head, in the fashion of a crown, with speckled feathers, pleasant to behold ...'.

The growth of ocean trade brought many shipwrecks:

'Sometimes are cast in by the Sea … Hogsheads of Wine and Brandie, all covered over with an innumerable plenty of these Creatures which they call Cleck-goose [barnacles] …'. A century later, the lighthouse engineer Robert Stevenson wrote: 'The author has actually seen a park paled round chiefly with cedar-wood and mahogany from the wreck of a Honduras-bound ship, and in one island, after the wreck of a ship laden with wine, the inhabitants have been known to take claret to their barley-meal porridge'.[3]

In the 20th century, the poet and naturalist Robert Rendall wandered the shores of Birsay, steeped in the study of rock pools and awed by their beauty:

Life spills her myriad forms before our gaze

In tiny treasures – bright anemones,

Worms, star-fish, crabs and little fish that leap

Across the pools. Look how storm waves heap

A fringe of shells along these sandy bays,

And how on golden bladderweed that sways

With rhythmic motion periwinkles creep ...

Among Robert's friends were fishermen and beachcombers:

I knew a Birsay man, an old beachcomber

Who gathered driftwood, holding it in scorn

To lose one piece. His thoughts, like those of Homer,

Compassed the ocean ...[4]

Over many centuries the sea provided a bounty of timber. Harry Berry wrote of his friend Sam Dick on the island of Hoy: '... his hobby was building small boats into which he incorporated pieces of seasoned driftwood salvaged from the Pentland Firth shore which he combed frequently. I remember the day he picked up the piece of beautifully grained Balkan maple and how, he had told me, he would make a fiddle from it. Eventually, he gave up his boatbuilding and devoted all his spare time to the making of his fiddle and, after six months of infinitely careful work, it was completed.

I had seen it only the once, when it was finished. It hung above his bench, perfect. The lovely grain of the Balkan maple body, the slender neck of Cuban mahogany, cut from some choice piece of some wrecked ship's furniture, the scroll exquisitely carved, the ebony fingerboard that had once been some mariner's chart ruler – the driftwood fiddle.'[5]

From time to time, far-travelled messages are found on these shores. They are contained in bottles, and occasionally in tiny St Kilda "mailboats", first used as a cry for help in times of famine, and now a ritual among the working parties that visit the island. In Orkney, many maintain the tradition of collecting "groatie buckies", the rare and beautiful miniature cowrie shells to be found at certain places around the shores.

Several decades ago, a huge deck cargo of timber found its way to the North Isles, causing great excitement and activity in securing it above the high water mark. Today, in the age of the container ship, there is much less timber to be found, but it has been replaced by tangled heaps of ropes, nets and bottles, all of indestructible plastic. This has bred annual expeditions to "Bag the Bruck", but one man's bruck is another man's treasure, and a walk around the shore can still reveal the useful, the beautiful and the mysterious.

The photographer Keith Allardyce first came to Orkney as a young man, absorbing the character of island life as a local assistant lighthouse keeper and an RSPB summer warden. He is well known for his best-selling books on the northern lighthouses, *At Scotland's Edge* and *Scotland's Edge Revisited*: and his photographic record of Stromness in the book *Sea Haven*. In the following pages, Keith Allardyce vividly records the stories and finds of contemporary beachcombers, revealing treasures still to be gained in an ancient and respected tradition.

Bryce Wilson

1 *Waith and Wrack*, Duncan John Robertson (1860-1941) from *An Anthology of Orkney Verse*, ed. Ernest W. Marwick, The Kirkwall Press, 1949.

2 *A Description of the Isles of Orkney*, Rev. James Wallace (d.1688), late minister of Kirkwall. John Reid, Edinburgh, 1693.

3 *Journal of Robert Stevenson*, Engineer to the Northern Lighthouse Board.

4 *Shore Poems (In the Ebb and Old Jeems)*, Robert Rendall (1898-1967), The Kirkwall Press, 1957.

5 *The Driftwood Fiddle, Tales from Harry Berry*, The Orkney Press, 1990.

FOREWORD by Tom Muir

I was born in a small house by the side of Inganess Bay in the parish of Tankerness in the East Mainland of Orkney. As a child, the sea was a constant presence in my life, and I spent many happy hours searching for treasures among the rocks and seaweed. In those days (the late 1960s, early 1970s) the shore was still used as a place to dump your unwanted rubbish; everything from broken household ornaments to wrecked cars went "over the banks" to be slowly ground down to nothing by the relentless tide. Even the roof over my head may have come from the sea (I suspect), as the red Canadian pine of the rafters resembled the wood that had come ashore off the *Loch Maddy*, an extremely unlucky ship whose remains lie in the bay below our house. She was part of the convoy HX.19, which was sailing from Vancouver to Leith with a cargo of wheat, wood and aircraft parts when on the 21st February, 1940, she was torpedoed off Copinsay by U-57, but remained afloat and was taken in tow. The following day she was again torpedoed, this time by the U-23, and broke in two, her bow section sinking in deep water. The stern was towed into the shallow waters of Inganess Bay, where she lay in 12 m of water. As a boy, I remember the buoy that marked the wreck, but in the late 1970s it was blasted and more wood was salvaged. The explosions dispersed the wreck, which meant that it no longer needed to be marked. The wood was badly eaten by marine worm, rendering it apparently useless. However, that same worm-eaten wood was used in fitting out the décor of the Bothy Bar in Kirkwall, giving it an antique feel. It was on that same shore where, one happy day, I found a book, *How to be Topp*, the hilarious story of an unconventional pupil of a boarding school called Nigel Molesworth. Written by Geoffrey Willans, and illustrated by Ronald Searle (of St Trinian's fame), it was gleefully read and reread many times. That particular book is long gone, but I still have a copy of it at home to this day.

A Second World War blockship, the oil tanker *Juniata* had been beached at the end of the bay below the airport, with the intention of being broken up for scrap but with falling scrap prices it was abandoned to slowly rust away. This wreck, known to me only as "the owld ship", seemed slightly sinister and scary to my young and rather over-active imagination. The sea came into childhood stories, too. My mother came from Westray, and knew the remarkable tale of a small boy who was the only survivor of a shipwreck there in the 1730s. A piece of wood bearing the name of the ship's port of registration, Archangel, was the only clue to the origin of this Russian ship, so the boy, who was adopted by the man who found him, was given the name Archie Angel.

Keith Allardyce has a long association with Orkney and a keen interest in its maritime history. He is a creative and original thinker who has had the vision to bring the past and the present together in books like Sea *Haven* and *Found*. With the eye of an artist, he brings out the best in the subject being photographed, whether it is a piece of flotsam washed ashore on an Orkney beach or the person who found it. The very concept of this book is in itself a good example of his unique style; what has been found washed up and what is the story that lies behind it? It gives us a fascinating glimpse of the state of the world in the early 21st century; a window on the social history of these islands through what has been washed ashore or dumped there. It also lets us look back at our maritime heritage, as many pieces shown here were found long ago. Imagine if this book had been written in 1912 instead of 2012. What would have been found washed ashore then? Steamships had taken over from sail, but losses like the *Titanic* showed that it was not mankind who ruled over the sea. The Orkney shore would have been strewn with things from all over the globe as the trade network stretched from Britain to the shores of its far flung-empire over the seas. Tramp steamers were not the safest vessels, and nearly every week *The Orcadian* newspaper carried stories of the loss of ships in heavy seas. Imagine, again, if this book was written in 1812 when Britain was at war with both America and Napoleonic France. Ships relied on sails, putting them at the mercy of the elements. What would you find after a gale if you walked the Orkney shoreline? Maybe things that nobody would ever want to see.

This is a book to treasure and to keep. It will be read and reread with as much glee as that Molesworth book of my childhood, not just by our generation but by generations of people yet to be born. This is not just a book that captures a moment in our time, but a book that is truly timeless. It belongs to the past, the present, and the future, and will bring joy to everyone whose lives it touches.

CONTENTS

DONNY SUTHERLAND and OIL PAINTING, Longhope

Following my chat with producer/presenter Dave Gray on Radio Orkney about the *Found* project, a letter arrived from Ann Sutherland of Longhope. Ann described an oil painting of a ship under sail which her husband found in the 1970s in a tiny geo on the south side of their farm.

The following summer I visited Ann and her husband Donny. 'It was floating in the sea at Rothie Geo, just about six feet from the shore', said Donny. And there was the painting, brought out from their spare room where it has been stored for the last thirty-five years. We were unable to read the artist's signature.

'It would need a frame if we were to hang it on the wall, and we haven't got it framed', said Ann.

'And I found a yellow submarine a few years ago', said Donny.

'An oil painting AND a yellow submarine!' I said. 'And anything else?'

'Flip-flops!' said Ann, 'So many flip-flops, but never a matching pair.'

The yellow submarine was only three or four feet in length, and Donny thought it was used for cable inspections. After he returned it to the company it belonged to, they sent him a twenty-pounds reward.

I went with Donny down to the geo where he'd found the painting. We walked across fields of rough pasture, the painting secure under my arm. Before us stretched a wide Pentland Firth panorama, from South Ronaldsay in the east, ahead the silhouettes of Swona and Stroma and along Scotland's north coast, and finally west to the dark hills of Hoy.

We took a steep path down into Rothie Geo, to its tiny beach with an assortment of driftwood, and bleached whale vertebrae, and a dolphin skull. 'This is where I found it', said Donny. The bright Orkney light merged the sea and the sky for a perfectly clear background, and I asked Donny to hold the painting up for a portrait.

ELMA SEATTER and *JOHANNA THORDEN* BUCKET, Orphir

'It's not easy to pass through a narrow piece of water like the Pentland Firth when you've just steamed across three thousand miles of the wide Atlantic Ocean', said Cyril Annal, as he told me about the tragedy of the mv *Johanna Thorden*. She was a 5,500 ton vessel carrying forty-five people, and a very valuable cargo of copper bars, steel plate, engines, compressors and tobacco.

On her maiden voyage from New York to Gothenburg, Sweden, mv *Johanna Thorden* entered the Pentland Firth in heavy snow on 12th January, 1937. 'The captain mistook the newly-built Tor Ness Lighthouse on Hoy with the Swona light', continued Cyril. 'His charts weren't up to date. The vessel struck the south end of Swona.

'The first my uncle, James Rosie, knew of the wreck was shortly after he'd been to check the Swona boats, early in the morning', said Cyril. On seeing wreckage floating past the entrance to the little harbour, James quickly alerted the other eight islanders. But there was nothing they could do to help.

'The *Johanna Thorden* broke in two soon after she struck the rock', said Cyril, 'and she sank in deep water. Only eight people survived, coming ashore below St. Peter's Kirk on South Ronaldsay after drifting in one of the ship's lifeboats'.

With her brothers, Kenny and Jimmy, Elma Seatter holds a wooden bucket, a handsome and rare memorial to the tragedy of the *Johanna Thorden*. They are standing on the Scapa Flow shore at New House, Orphir, where the bucket was found by their father, Harry Johnston, soon after the ship was wrecked.

IAN COOPER and MASK, Stronsay

The fertile fields of Midgarth on Stronsay's north-west corner face the small island of Linga Holm, just off shore, and beyond, the islands of Eday, Egilsay and Rousay, and many holms and skerries. Ian Cooper farms this part of Stronsay, and it was along the shore of Linga Sound where he found this African mask.

'I turned over something half-hidden in the seaweed with my foot', said Ian, 'and this face suddenly appeared. It must be over thirty years ago now'. 'We don't have it hanging on the wall', said Margaret, Ian's wife. 'You don't know what it means, or what it was for'.

Tom Muir of The Orkney Museum made an enquiry on my behalf at the National Museum of Scotland, and he received this reply from Chantal Knowles, the Principal Curator of the Oceania, Americas and Africa Department at Edinburgh's National Museum of Scotland:

'...an African mask. Probably tourist art piece - may be even export art and of no particular region, more a generic African style. These types of things are probably exported in vast numbers for home/interior shops across Europe and North America'.

This response might take a bit of the mystery and romance out of this find, but it is still a beautiful object.

ETHEL MOWATT and PALM FROND CARVING, Graemsay

Ethel Mowatt has lived all her life in Graemsay, where she ran a croft with Ronnie, her late husband. Ronnie was also an occasional lighthouse keeper at Graemsay's Hoy Low Lighthouse. Often on the Graemsay shore, Ethel discovered around 1980, a palm frond with a fierce-looking face carved on the thick, woody base.

The face on the frond looks like the expression made by a New Zealand All Black or Kiwi rugby player during a pre-match ritual war-dance.

An inquiry with Chantal Knowles, at Edinburgh's National Museum of Scotland, suggested: '...it may be Indonesian, as they export furniture and wood carvings internationally, but I can't say I have ever seen anything even reminiscent of it!'

IAN SCOTT and WHALE SKULL, North Ronaldsay

In the silent gloom of the byre at Antabreck, his croft on North Ronaldsay, Ian Scott stands beside a plaster of Paris model called *Sea Forms*; it is a sculpture which was inspired by the whale skull he is holding in this portrait.

This rare find was picked up by Ian at Nouster Bay, near the island jetty, in the 1970s. Unfortunately, on lifting the skull out of the safety of its box, lined with tissue and newspaper, this fragile and precious object fell apart into four pieces, as shown here.

Having studied Art and Sculpture at Aberdeen's Gray's School of Art, Ian Scott graduated in 1962 and returned to North Ronaldsay to work on the Antabreck croft, to pursue creel fishing, and to continue painting and sculpting. *Sea Forms* was made in 1989, and Ian hopes that one day it could be cast in bronze. He is already well known for his bronzes commemorating the lifeboatmen who lost their lives so tragically at Longhope and Fraserburgh.

JOHNNY MEIL and SCAPA FLOW FINDS, St. Ola

Orkney's father of all beachcombers must be Johnny Meil. With his wife, Isobel, he lives at Dyke End near Kirkwall, in a house perched high above Scapa Bay with commanding views across the eastern stretches of Scapa Flow, the *Royal Oak* buoy, and over to Hoxa Head on South Ronaldsay.

'I started beachcombing at the age of three', Johnny said, 'along the shores of Scapa Flow, and I've lived here all my life'. In an enormous shed crammed with naval history, I photographed Johnny (in 2011, when he was 87) with two of his remarkable finds. We decided to include a former bench top from the famous *Royal Oak*. He picked it up off the shore soon after the ship was torpedoed in 1939. Johnny is holding a block of wood with such a complicated arrangement of cubicles within it that it could only be an apprentice-piece, probably from a new recruit to a ship's carpentry workshop.

Amongst this fascinating collection, Johnny told me, were '...two long boat hooks from the steam pinnace of HMS *Ramiliese* (a battleship which saw action in both world wars), a flag-staff and flag from a naval vessel which plied between Scapa and Lyness, a beautiful 15ft spoon-bladed Royal Navy oar with two copper bands, tobacco tins, navy hats, lovely pieces of hardwood from ships' carpenters' shops, submarine markers, a 20ft long torpedo, part of the decking of the *Royal Oak* pinnace, and many sturdy wooden boxes ('perfect for making hen-houses')'.

7

THE DODMAN FAMILY and FINDS, Papa Westray

On moving to Papay (as Papa Westray is usually known), the Dodman family gave this small island a significant population boost. Established on Papay's north-east coast since 2001, the family is six in number in an island population of about 70 today.

Scanning the Papay landscape through my binoculars on a breezy, showery day, I first noticed their house, Hundland, nestled close by the shore of a little bay, North Wick. Drawn towards it, I could make out a typical Orkney long house, stone-built and with a fine flagstone roof. The house looked mysterious through the rainy haze.

Papay is full of surprises. Along the shore of North Wick we met Andre, who was here on holiday, and he told us, just before slipping into the sea, 'I swim with the seals every day.' Half a dozen seals were hauled out on a reef not 50 yards away. 'Back in Paris', he called from the water, 'I allow people to fly... I work in a theatre'.

A little later we were given a warm welcome at Hundland after we knocked on the door to enquire about any sea finds. Over cups of tea, we heard about life on Papay, and about their earlier life in South Africa, and conservation, and were shown around their home.

There were small finds from the shore below Hundland all over the house, and in the bathroom there were two massive flagstones, carried from the shore and beautifully fashioned to make two sides of the room. In the corner, appropriately, stood a weathered depth measure from the Westray pier.

Outside the house, sea finds lay all around: whalebones, driftwood, fishing boxes and baskets, a dog bowl, a football, and netting and rope. For a portrait in front of Hundland, we chose a melamine plate and a small soft toy, and an orange pole from the mv *Stella Rigel*, a Dutch cargo ship.

8

HELEN SWANNEY and *MIM* BENCH, North Ronaldsay

The smell of remoteness and the sea hangs in the North Ronaldsay air. This is an island which stands apart from the rest of Orkney. Scattered around the coast, and around many of the island crofts and houses, lie many objects from the shore.

At Trebb, near the airstrip, Helen Swanney has presided over her shop for half a century, and keeps an eye on things as coordinator when the Loganair planes arrive and depart. Helen's late husband, Ronnie, worked on the laird's home farm. All around Trebb are many trophies from the sea. A large lower mandible from a cetacean stands in a window sill, and beside a redundant petrol pump a three-stone bale of Indian rubber is slumped, having been washed ashore in 1937 after the *Johanna Thorden* sank off Swona.

'And this is a capstan from the *Garibaldi*', Helen told me, as she showed me around. In the window of an outhouse 'that's some sort of housing, with four portholes, from the *Hansi*; it went down in 1939. I remember, because it was wrecked near my aunt's house, and I was always there when I was young. There were many bales of paper pulp washed ashore from the ship', continued Helen, 'but I did find a big tin of varnish. That was the most useful thing.'

Helen is photographed here, sitting in her shop on a bench from the ms *Mim*, a ship which was wrecked on Reefdyke, just a week before ss *Hansi* went down, in November 1939. (The crews of both ships were all rescued.) 'The hens ate well for a long time after the *Mim* went down', said Helen, as the cargo had been a consignment of wheat, bound for Norway from Fremantle, Western Australia.

9

JOHN O'WESTNESS and part of *MIM* HATCH-COVER, North Ronaldsay

Known to everyone on North Ronaldsay as John o'Westness, (because he was born at a croft called Westness), John has lived at Trebb with his sister for over ten years. Retired now from crofting, he is photographed here holding a short section of timber, an off-cut, which was originally a twelve-foot length of a ship's hatch-cover.

'This is about all I have left of my hatch-covers', said John. 'Hundreds of them came ashore around North Ronaldsay during the war. I've made a few shed roofs with these', he said. The hatch-covers were a rare and valuable resource, and have been used in many island buildings, after ms *Mim* was wrecked on Reefdyke.

Bound for Norway from Fremantle with a cargo of wheat, the *Mim* was suspected of being German when spotted by the crew of a British frigate. After putting a British officer on board, the *Mim* was ordered to sail for Scapa Flow. On the way, however, on 1st November, 1937, the ship struck Reefdyke in heavy seas. Twenty-four hours later, the Stromness lifeboat reached the *Mim* and twenty-two of the crew were taken to Kirkwall. The remaining ten had made it to shore in the ship's tender.

BERTIE & ALICE THOMSON with DYEWOOD and CHAIR, North Ronaldsay

The astonishing voyage of the Swedish East Indiaman *Svecia*, a 600-ton armed merchantman, began in early summer 1740, in Bengal, where she was loaded with a vast and valuable cargo of dyewood, saltpetre, silks and muslins, and many chests crammed with the personal belongings of over one hundred passengers and crew.

After a journey of around 10,000 nautical miles, the *Svecia* had only the North Sea to cross before making it home to Gothenburg in Sweden. But as she was heading between Fair Isle and Orkney, *Svecia* was driven south by gales and caught in the strong tides that drove her onto Reefdyke, a notorious rock one and a half miles off North Ronaldsay.

Despite being so close to the shore, and in clear view, there was no attempt made by local men to rescue the crew. For three days the stricken ship took a terrible pounding from the sea. Survivors later denounced the islanders as barbaric for not helping them. But the simple truth was that none of the small island boats could have survived in the rough seas if they had tried to reach the stranded ship. The fact that the *Svecia's* two lifeboats were swept away north when they were launched showed the power of the sea at the time.

One of the lifeboats was wrecked on Fair Isle and 31 survivors scrambled ashore. The remaining officers and crew cut down the topmasts, and with the use of rigging they made a raft. Hoping to make landfall on either North Ronaldsay or Sanday, they were also carried away north. At Dennis Head, where the Old Beacon now stands, the raft was swept with a huge wave, washing the men into the sea, never to be seen again.

By ripping up part of the deck, the last 24 on board made another raft, but only 13 of them made it ashore alive.

Of the 104 people thought to be on board at the beginning of the *Svecia's* journey, only 44 survived.

The dyewood, part of the *Svecia's* cargo, doesn't float, and has been rolling onto North Ronaldsay's beaches ever since the loss of the ship. It was from a type of Indian sandalwood, and is now extinct.

Bertie and Alice Thomson were photographed at home with pieces of the 'Svecia' dyewood and a chair from the *Mim*.

11

ALBION CROCKERY and BRICKS, Graemsay

On Saturday, 21st December, 1865, the 1,225 ton fully-rigged ship the *Albion*, left Liverpool bound for New York. On board were 43 emigrants from Germany, a crew of 24, and a cargo including steel, iron, tin, crockery and bricks.

Soon after setting sail, the *Albion* was in trouble. Persistent westerly gales drove her back towards the coast, where she nearly struck St.Kilda. Forced even further north, she rounded St. John's Head on Hoy and entered Hoy Sound where the captain dropped anchor.

It was now 1st January, 1866. Two Orkney pilots reached the ship to advise the captain to cut her cables and seek shelter. He didn't, (the crew were utterly exhausted by this time), and the ship started to drag her anchors in the storm. Then, shortly after noon, the ship struck Graemsay's Point of Oxan, just below Hoy Low Lighthouse.

Being New Year's Day, the Graemsay men were all playing a game of mass football called The Ba', as was the custom in most of the islands and parishes. The game was stopped, of course, as soon as it was clear the *Albion* was heading for disaster.

Several boats from Graemsay and the surrounding area went out attempting to aid the ship, including the paddle-steamer *Royal Mail*. One boat belonged to Joseph Mowat of The Hill on Graemsay, and he managed to take on board nine people. But, in the confusion and chaos, his boat was overturned by one of the mailboat's paddles, and he was drowned along with all his passengers.

The *Orkney Herald* described the tragedy in detail under the headline "A Melancholy Shipwreck", and reported : 'Before 5 o'clock the once stately *Albion* was completely broken up, and the Graemsay beach was strewn with pieces of the wreck and portions of the cargo'. The fifty-seven survivors were given food and shelter at Graemsay's two lighthouses until they had the opportunity to cross to Stromness and make their way south.

Two chests of *Albion* crockery are believed to have been buried on Graemsay to avoid detection. One was discovered in the 1950s. Salvaged wallpaper from the cargo was auctioned, and for a while hung over the pews of the United Presbyterian Church in Stromness to dry.

At the Point of Oxan, where the *Albion* was wrecked, there are still traces of the tragedy to be found. Amongst the stones, there are pieces of the ship's china, some in the shape of various fruits, which served as lid handles. There are also the remains of the cargo of bricks, now sea-worn and rounded, but still brightly speckled in yellow, orange, black and cream.

12

BRYCE WILSON and WHALEBONE, Graemsay

In the early hours of Saturday 14th July, 1906, a massive sperm whale (some say 90 feet in length), came ashore on Graemsay. The species is the largest of the toothed whales - Herman Melville's fictional Moby Dick was a sperm whale - and it's second only to the biggest animal which has ever lived, the blue whale.

A fishing boat had originally dragged the whale into Stromness Harbour, but nobody wanted it. The body was then towed beyond Hoy Sound and released into the Atlantic, only to be carried back on the tide, and ending up tightly lodged in Geomarion on Graemsay's west side.

Word quickly spread across the island when the great whale made its appearance. Graemsay's population at this time was around two hundred people, and all day a steady trickle of islanders made their way to view the unfortunate creature. Photographs were taken while people balanced on its head or posed below its huge flanks. But, meanwhile, the whale was stealing someone's thunder. Having had their curiosity satisfied, the sightseers then called at Windywalls, just above the shore, to give their blessings (albeit a little late) to a baby girl who was born on the very morning that the whale found its way to Geomarion. That baby girl was Bryce Wilson's mother.

This portrait of Bryce was taken at Geomarion with a souvenir from the whale. It is one of the vertebrae, weighs two stones and two pounds, and was used for many years as a milking stool at Netherhoose, one of the island's crofts. A few more bones, and no doubt a few teeth too, were salvaged from the whale that summer in 1906, but soon the smell of the rotting carcass meant that it had to be removed.

The Orkney Customs authority engaged Captain Harcus of the *Hoy Head* to drag the body out of Geomarion and out of harm's way, when it was released somewhere off the Brough of Birsay, never to be seen again.

13

JACKO & DOREEN LINKLATER and St. KILDA MAILBOAT, Quoyloo

St Kilda is a small group of islands, west of the Outer Hebrides, and about 100 miles west of the Scottish mainland. Remarkably, the main island, Hirta, once supported well over 160 people, but in 1930, with a dwindling and ageing population, it was evacuated at the request of the islanders.

For many years, until a post office was installed to cater for the tourist trade in 1898, the method used by St.Kildans to communicate with the outside world was to put small model boats with cargoes of letters into the sea, in the hope that someone on a distant shore would find them.

The mailboat was the brainchild of a freelance Scottish journalist and artist, John Sands (1826 - 1900). During his visit to St.Kilda to excavate an iron-age fort, nine Austrian sailors were shipwrecked and stranded on the islands, causing the local food supplies to run dangerously low. Sands dispatched a mailboat into the sea with a message requesting food supplies for St.Kilda. The mailboat was found in Birsay, about a month later, and soon a well-stocked ship with produce from the farms and fishing ports of Orkney arrived in Hirta's Village Bay.

An Army base was established on St Kilda in 1957. Occasionally, someone stationed there would continue the mailboat tradition. Then the late Jacko Linklater, on a Birsay beach in 1982, picked up a model battleship and found a number of stamped and addressed letters inside a compartment, which he duly posted.

Jacko found a few other St Kilda mailboats in his time along the Birsay shore. And also two dugout canoes, possibly from the Caribbean, fishing buoys from St John's, Newfoundland, a few messages in bottles, oars, bamboo fishing rods, herring barrels and planks, hundreds of pit props, and something still used today in his house, a floor brush.

Jacko had many interests, along with his crofting and fishing. He was a talented musician, and played the accordian to accompany his father on the fiddle. This duo, the Hoosegarth Band, was popular in Orkney, playing at hundreds of local weddings, dances and harvest festivals.

Jacko passed away on Wednesday, 28th December, 2011.

14

LINSEY DREVER and FIGURINE, Deerness

For well over two centuries, until the Second World War, smoke billowed from hundreds of kelp-burning pits all around the Orkney shore, where great quantities of tangles were burnt during the summer months. The burning produced a valuable ash, rich in potash and soda, vital ingredients in the glass and soap industries of the south.

In 1931, during the Great Depression, a Deerness farmer, Magnus Wylie, found this Japanese figurine on the shore while he was gathering tangles for his kelp-burning pits, between Newark and the Point of Ayre.

In this portrait, with her husband Graham and son Joe, Magnus Wylie's grand-daughter Linsey Drever holds the figurine on the Deerness shore where it was found over eighty years earlier. 'Did you know, I'm descended from someone who was found on the shore?' asked Graham. I asked what he meant, and Graham told me the story, well known in Orkney, which began with a shipwreck in the 1730s in Westray.

One stormy night in the island, a ship struck the Aikerness rocks and was soon broken to pieces. It appeared that there were no survivors of the disaster. In the morning light, many bodies could be seen, strewn along the shore. But then one of the islanders, a man called Rendall of Seaquoy, noticed a woman lying still and cold, and tied to her for his protection was a small boy. He moved a little, and gave a faint cry. So the man gently lifted him up, and took him home to his wife and the warmth of the fireside. Soon the boy revived, and he became another member of the family.

Where the boy came from, or what his name was, no one knew. A piece of the wreck which came ashore with the boy had the name of its port of registration written on it, Archangel, in Russia. The boy was then forever called Archie Angel. He kept the name, and grew up in Westray, and had children. Until the late 19th century, there were Angels in Westray, and though the name Angel has died out in Orkney, there are today descendants of Archie all over the islands.

15

JIMMY MOWAT with LIFEBELT and GREEN DOOR, Hoy

About seventy years of battering by Orkney's notorious weather has produced the best character a shed could have. And what is also special is the green door of this shed. The door came off the Hull trawler *Siherite*, after it was wrecked in dense fog on Rora Head, near The Old Man of Hoy on 16th March, 1936.

'My father and grandfather went aboard the wreck and took as many fittings as they could - bits of brass, and this green door. And the porthole. They came from the bridge', Jimmy explained. 'I'm from a fishing family and lived in Rackwick at the time - just round the corner from Rora Head. It's where I lived until I was five', continued Jimmy, 'in a house called The Moss'.

The name The Moss rang a bell with me, and I looked it up in George Mackay Brown's *An Orkney Tapestry*. There was The Moss in the sixth verse of the starkly simple elegy to Rackwick, after the village became deserted except for one man in the 1950s :

The Moss is a tumble of stones.

That black stone

Is the stone where the hearth fire stood.

The lifebelt which Jimmy is holding was found on the shore near his home in Brims. It was washed ashore soon after the German trawler *Thumfisch* struck one of the Pentland Skerries on 31st March, 1975. The boat had a crew of twenty, and all survived. Three of the crew were picked up by the Longhope Lifeboat, while the rest drifted in life rafts on to Swona, where they found shelter in one of the island's deserted houses. It was a house I knew well, Norhead, as I had occasionally stayed in it myself the previous year. I was astonished to see that remote house again on ITV's *News at Ten,* with the newscaster Sandy Gall telling the story of the Swona rescue.

16

IAN & MARGARET COOPER'S GREYHOUND TILLER

Execution Dock, London's gallows by the Thames in Wapping, was specially reserved for hanging pirates and other maritime offenders until the 1830s. It was here, on 11th June, 1725, that Orkney's most notorious pirate, John Gow, ended his life at the age of 28.

Earlier that same year, Gow and his crew, in their ship *The George*, (the third name given to the ship in Gow's hands), were in Orkney waters. Knowing Orkney well, Gow's intention was to raid a few lairds' houses. Easy pickings for a change, he thought.

'During a fairly bungled attack on the Hall of Clestrain, legend has it that Gow and his men missed some valuables hidden amongst some feathers in an upstairs room, and even under the skirt of the laird's daughter, who calmly sat still during the raid. The hapless pirate's next target was Carrick House in Eday. But with a combination of incompetence and the awkward currents of Calf Sound, Gow's ship became stranded on the Calf of Eday. Remarkably, this gave an opportunity for James Fea of Carrick House to organise, through a mixture of elaborate trickery, negotiation and deception, the capture of the pirates.

They were held in Eday until support came three days later with the arrival of HMS *Greyhound*, a 20-gun Naval vessel which specialised in hunting pirates. The prisoners were soon delivered to Edinburgh, and ultimately to London and a trial at the Old Bailey. During all the drama of those few days around Calf Sound, one item, the beautifully carved tiller from the launch of HMS *Greyhound*, was left behind, either lost or stolen.

The tiller then somehow found its way to Midgarth in Stronsay, where it is lovingly treasured to this day. Ian Cooper, who farms Midgarth, explained 'At the time of John Gow's capture, James Fea lived at Carrick House on Eday where Gow was held. A cousin of Eday's James Fea, also called James Fea, farmed at Clestran not half a mile from Midgarth. Did one cousin give it to another, and someone gave it to their neighbour at Midgarth? The tiller was in this house when my grandfather moved here in 1948'.

James Fea of Carrick House ended up seriously out of pocket over the Gow affair, fighting counter-claims on his reward money through the London courts.

17

JIMMY & MARGARET MOAR with BUOY and RAT TRAPS, Hoy

Below the steep slopes of Hoy's mighty Ward Hill, and overlooking Scapa Flow, is the sturdy white cottage of Jimmy and Margaret Moar. Their home is a treasure chest of sea finds.

Outside their front door stand two five-feet-long whalers' harpoons 'from a wreck on the Kirk Beach near Stromness', says Jimmy. There is also a large blue plastic drum from the shore, with pieces cut out of it to make a house for their two cats.

A large pile of colourful buoys fills a corner of the garden, mostly gathered from the tiny beaches below the sheer Hoy cliffs - Little Rackwick and the towering St. John's Head. Jimmy always had a small dinghy with his lobster boat whenever he was out creeling, to allow him to manoeuvre into a difficult landing.

'Rackwick was our main haunt', said Margaret. 'It was where we found this strange buoy with an antenna, and these American rat traps. But when Hutch moved to Rackwick, it all changed. He always made sure he was out along the shore very early, long before we got there.'

'Our best find', Jimmy continued, by now thoroughly warmed up to his beachcombing memories 'was a huge inflatable life raft. It was a typical bright orange colour, it was a dome with a hatch. Big enough for twelve people. Well, we'd take it on holiday and use it as a tent! The first time we used it was at a campsite in England. We had a trailer to carry it. Then we began to set it up - and soon a crowd gathered round as we started pumping. This thing grew and grew. It dwarfed everything else on the campsite. It was like something from another planet...'

18

COLIN KIRKPATRICK and DRIFTWOOD ART, Stromness

Studying in the 1990s at Gray's School of Art, Aberdeen, Colin Kirkpatrick was a pupil of one of Orkney's most respected teachers, Sylvia Wishart. 'She was always encouraging, always positive.... She had an Orcadian way of giving criticism in a gentle, constructive way', he said. 'And her humour too. She told one student, to suggest he should work harder, 'You need a squib putting down your knickers!".

Specialising in working with driftwood until recently, Colin has always picked up driftwood, and he continued to use it through art school. Then later, during a time as a lighthouse keeper at Copinsay and Fair Isle South, his need to scour remote beaches was reinforced.

Now living with his family at Outertown, with views over Hoy Sound and the Atlantic, 'I live only a stone's throw from Billia Croo, a great wee beach, and Breckness, another beach which seems to attract a fair amount of driftwood', said Colin.

19

RONALD FOUBISTER with FISH BASKETS and SHED of FINDS, Deerness

The last attending boatmen to Copinsay Lighthouse until its automation in 1991 were brothers Ronald and Davey Foubister. Regular fortnightly trips were made from Newark Bay in their yole, supplying the keepers with gas cylinders, groceries and newspapers.

I first met them when, as an RSPB summer warden in 1974, I made occasional visits to Copinsay, a half-hour journey, and stayed in the island's farmhouse. There were people around this part of Deerness at the time. Apart from Ronald and Davey, and myself, there were three keepers at the lighthouse, and in a tiny, wooden house built into the dunes at the south end of Newark Bay lived Willie Johnston, who had returned to his homeland after living in Australia for many years. Lengths of old fishing nets strung along a line of posts gave shelter from the east winds. Over time, the house was covered with sand, blown from the beach, and only the wooden frame of a porch gave a clue that anyone was living here.

Finding driftwood on the shore for his stove and for repairs to his house, Willie also collected tangles throughout the year, often with Ronald, for the alginate industry. A ship from Kelco Ltd in the Western Isles made annual visits to many parts of Orkney to collect and pay for the harvest.

Ronald is photographed here near his house in Deerness.

20

CHRIS ROWELL and STARFISH, Hoy

The strange and eerie landscape of Lyness, its derelict buildings and huge corrugated iron sheds, a house which was once the foyer to a garrison theatre, and brick chimney-stacks standing isolated in small fields, was once the hub of naval operations in Orkney during the Second World War.

On a hillside above the chaos stands a vast, gaunt ruin, which on a misty day seems reminiscent of some eccentric Moroccan castle. This extraordinary building was once the Naval Headquarters and Communications Centre, the very heart of Orkney's wartime activity.

Far below, nestled amongst willows and shrubs near Mill Bay, sits a small house, the home of Chris Rowell. 'If there's one person you have to visit, then it has to be Chris...' insisted Effie of Harray, an old friend. 'I won't tell you any more', she said, 'just see for yourself.'

The signs of a dedicated beachcomber lie all around the garden. Sea finds festoon every bush, path side, every ledge, everywhere. Radio 3 plays Mahler from an outside loud-speaker. Seal skulls, buoys and driftwood, large handles from army tea cups and mugs, plastic dolls, shells... all from the shore, and all given a new life.

And inside the house, too, an Aladdin's cave of sea-treasure. A visitors' book lists many comments when Chris ran a unique cafe here a few years ago : 'An amazing, magical place. Mysteries and joy tucked away in every corner', wrote one visitor. Another wrote 'Thank you for one of the most bizarre house tours I've ever experienced.'

'Come and have a look at the bathroom', said Chris, and there, as I opened the door, lay the perfect setting for a portrait. Chris stood in the bath, and picked up an ornament, a dried starfish, and the photograph was taken.

21

CRISPIN & CHARLOTTE WORTHINGTON with DECK BRUSH and COCONUT, Stromness

On Baikie's Pier, named after the boat builder, Edward Baikie, who operated here building yoles and dinghies early last century, Crispin and Charlotte Worthington are photographed with two finds from the shore below them, a deck brush and a coconut.

Coconuts are regularly found on Orkney's beaches, '...and probably come from the West Indies', said Crispin. 'I found a small loggerhead turtle on the Orphir shore a few years ago', he continued, 'and it was only the second recorded for Orkney'. (On display in Stromness Museum is a large leatherback turtle , which was found in a net by fisherman Edwin Groat off Westray a few years ago.)

Until recently, Crispin took part in the RSPB's monthly beached bird survey, always walking a stretch of the coast at the full moon, to take advantage of the higher tides, recording the numbers and species of dead seabirds he found.

22

DOUGLAS HOLLICK and part of BLOCK and TACKLE, Stronsay

It's hard to imagine today that the quiet backwater of Whitehall in Stronsay was once a bustling boom town, centred on the former herring industry. A fleet of around three-hundred fishing boats revolved around the village every summer, while four thousand people followed to take employment in gutting, salting and packing the fish, mostly for European export.

The bonanza began in the 1880s, but half a century later it was all over. Before the beginning of the Second World War, the very last herring boat had delivered its catch. This industrial-scale fishing couldn't be sustained.

There are many fine houses lining the waterfront in Whitehall, standing as monuments to that wealthy era (while many of the harbour houses were also built on the profitable kelp industry of an earlier time, from around the early 19th century). Otherwise, the paraphernalia of the herring industry has all but disappeared. Except, that is, for a trace, here and there.

For over three decades, Douglas Hollick has farmed at Huip, on the north end of Stronsay. Occasionally on the shore for walks near his house, he recently found this wooden object. It's probably made of oak, it has a groove around the circumference, and has a bronze central plate. It is part of a block and tackle - a simple but effective lifting device. I can imagine it might have been used to hoist many thousands of baskets, laden with herring, onto a busy Stronsay quayside; or it might have raised the billowing sails of an old Aberdeen trawler bound again and again for the herring shoals of the North Sea and the Atlantic Ocean.

23 SHIP'S TIMBER SCRATCHING-POST, Stronsay

'We've got thirty Charolais Cross cattle in that field, all colours', says Douglas Hollick, talking about the field in this photograph. 'It's called Stave on an old map of the farm', he adds – named after this huge piece of timber which one of Douglas's cattle is scratching itself against, perhaps. 'It looks like a piece of boat timber...could have been there for two hundred years', he says.

The scratching post is part of the furniture on Huip. It prevents the cattle from using other places to scratch, like fence posts. 'An earlier farmer had used a large stone as a strainer post, and when we were renewing some fencing we kept it standing. Someone had quarried it, and cut it, and carried it to the field with a horse and cart. We left it. A monument.'

Douglas and his wife came to Stronsay from Devon, '37 and a half years ago', he says. The farm is mixed, with grass and grain as well as the beef cattle. And they also run the little airstrip, attending to the twice-daily Loganair flights to Stronsay.

24

IAIN ASHMAN and FIRST WORLD WAR GERMAN CANDLE, Stromness

One of the most extraordinary events in naval history must be the famous scuttling of the German Fleet in Orkney shortly after the end of the First World War.

After an armistice was agreed on 11th November, 1918, between Germany and the Allies, seventy-four ships of the German Fleet were interned in Scapa Flow, Orkney's vast, sheltered, natural harbour.

Each ship was allowed to keep a skeleton crew, and all the vessels were under the command of one of Germany's top brass, Rear Admiral Ludwig von Reuter. But nine months had lapsed since the signing of the armistice, and with the unrealistic belief that Germany might not agree the peace terms and that war might recommence, and with the German Fleet in enemy waters, von Reuter ordered the scuttling.

The plan succeeded with the loss of fifty-two ships, while British guard ships managed to drag twenty-two ships to shallow water before they sank. The proud Rear Admiral was hailed a hero back in Germany for this final defiant act, although he remained a prisoner in Britain until 1920. Another of Germany's admirals was so impressed that he declared : 'The sinking of these ships has proved that the spirit of the fleet is not dead. This last act is true to the best traditions of the German navy'.

During the 1920s, an engineer, Ernest Cox, began an ambitious scheme to salvage many of the sunken ships. Initially, he bought twenty-six destroyers from the Admiralty for £250, and eventually all but seven of the fleet were raised from the sea bed.

Those remaining seven ships are now protected under the Ancient Monuments and Archaeological Areas Act. But, since the 1970s, small pieces of the ships' metalwork have occasionally been removed as it is a rare source of radiation-free steel, vital in the production of sensitive instruments in medical and space research.

Out for a walk along the Orphir shore, around the time of a scuttled ship being examined for its precious metal, Iain Ashman, a Stromness designer, found something which at first 'looked like a loo-roll holder - or a stick of dynamite !' It was a First World War German candle, and at some point in its history had been lit very briefly. Presumably dislodged from the ship's hold by a diver, the candle has imprinted along its side : 'KAISERLICHE WERFT', which translates as 'IMPERIALIST SHIPYARD'.

25

A *PENNSYLVANIA* BUCKET, STROMNESS MUSEUM, Stromness

One of Orkney's most well-remembered shipwrecks must be the ss *Pennsylvania*, a Danish freighter which had been heading for Copenhagen, but ran aground in dense fog on Swona in 1931.

The ship lay perfectly lined up on the shore below Rose Cottage, on Swona's west side, and the crew were safely disembarked and taken off the island the same day, leaving behind a cargo from the United States of breathtaking value and variety. On board were Fordson tractors and Dodge cars, tyres, fishing gear and copper bars, typewriters, tobacco and spark plugs, and boxes of shrouds and condoms. And thousands of packets of cornflakes. 'The hens laid marvellously on those', said Cyril Annal, as he told me the *Pennsylvania* story.

James Rosie, the island owner, members of his family and a small band of islanders from neighbouring Stroma, proceded to remove everything from the ship, and soon the holds were emptied.

Two months after the grounding, a raging westerly gale suddenly blew up, and within a few hours the great vessel had disappeared completely.

Traces of the *Pennsylvania* were later washed up around Orkney, including this wooden bucket, held here for this photograph by the Honorary Curator of Stromness Museum, Janette Park.

26

FRANK & SARA ZABRISKIE and WRECK TIMBER, Birsay

During a tour of the British Isles in 1984, Americans Frank and Sara Zabriskie visited Orkney to pursue their interest in history and archaeology. They were astonished at what they found, and fell in love with the islands - the place and the people. And two years later (Frank having retired from a career in astrophysics) they left their home in the foothills of the Appalachians, in Pennsylvania, to move into Harpsa, a house overlooking the Brough of Birsay and the wild Atlantic Ocean.

On their doorstep is one of the best parts of the Orkney coast for beachcombing, from the Brough to the Bay of Skaill. Soon, beachcombing became a passion. Most of their finds are stored in one of the Harpsa barns: huge pieces of the keel of a 64-ton fishing boat, the *Regent Bird*, which struck the Birsay shore in January 1995, fish boxes and ropes, a big gaffe, coconuts, and a bottle-message from a Faroese fishing crew '...we are at the Big Wheel on the Forotya Banks. We are seven men...'

Above a door to another outbuilding hangs a magnificent carving from the stern of an 18th century sailing ship. It is an elegant foliate design on a piece of wood measuring about eight feet by three feet. Bought in a job lot for a pound in a Burray sale, this came with an extraordinary variety of maritime miscellanea, including a huge ship's bellows, salvaged from the bottom of Scapa Flow.

And the "bruck truck", Frank and Sara's buggy, is an important part of their beachcombing trips. Itself made entirely from finds on the shore, it's a large plastic fish box, with four wheels made out of plastic floats, and a long piece of tow-rope, perfect for loading with driftwood.

27

TOM RENDALL and 'BANKS' WOOD, Westray

'I was born in the cabin of the barque *Emerald* a year before the outbreak of the Second World War', says Tom Rendall, as he shows me his workshop in Pierowall, a little village in the heart of Westray. The *Emerald* had been wrecked in 1879 on Aikerness Holm, just off Westray, and when this magnificent three-masted ship was brought into Pierowall Bay for breaking up, Tom's father bought the cabin to use as a dwelling house. It continued to be used for 123 years, and now stands, lovingly restored, in Westray's Wheeling Steen Gallery.

But it was Tom's skills with driftwood which brought me to his house, overlooking Pierowall Bay. 'This is my workshop', he told me, showing me a shed behind the house. 'And this is what I make from driftwood', as he held up his latest chair frame - destined to become another rare and coveted traditional straw-back chair.

The driftwood is no ordinary wood. It is what Tom calls "Banks" wood, perfect pine from the forests of northern Canada. 'This wood is occasionally washed up on Westray's shore, and it's free from sea-worms', Tom explained, 'and it turns up ready trimmed.'

I was just about to leave Tom, after having taken some portraits. 'This shed is made from the "Banks" wood, you know... ' he said, 'and my house is full of this wood too. I was building the house myself when a great number of the Canadian trees was washed up near here. I didn't need to spend anything at all on the timber.'

28

GEORGE COSTIE & SON with FLOAT, SKI and LIFEBELT, Westray

'It's my brother you'll need to speak to', said Alec Costie. 'He was always ahead of me whenever we went beachcombing in our youth. Always a bit keen...' Now retired from the fishing grounds of the north Atlantic and the North Sea, Alec lives by the Bay of Pierowall, with time on his hands for a change. I'm invited in for coffee and a few stories to keep me going.

'There was a Westray man back in the 1850s went down to the shore where 160 pilot whales had been beached, down at Cubbiegeo here. Many islanders were there to take their share of the unexpected bounty, the whale-meat. So this man came along too, after his wife had died early that morning. 'I'm surprised to see you here today', the laird said to him, as he went to the shore. "Well", he replied, "I can't afford to lose, in one day, both a wife AND a whale!"

'I found many sailors' caps on the shore during the war', said Alec. 'Blown off their heads? Lost sailors? I don't know. We wore them to school for a while. And my father found a beautiful brass telescope on the shore one day', he continued. 'A pocket of air in the casing had kept it afloat, but my father thought he could use it to repair part of the exhaust system on his boat'.

At the other side of the bay, I met Geordie and his son Heddle, just back from a day's creeling. Adventurous, this man Geordie. I saw him in his creel boat off Suleskerry, that remote outpost west of the Orkney Mainland, back in the late 1970s, when I was a lightkeeper there.

'Come into the house', said Geordie, where I met his wife, and saw his extraordinary collection of finds neatly on display. 'Here's a brick from the ss *Politician*', said Geordie, 'but I haven't got a bottle of the famous whisky'.

Back outside, we went over to Geordie's massive shed. The doors were slowly slid open to reveal enough found objects from the Orkney shores to fill a village hall jumble sale twice over. We raked through the collection and chose three completely different objects for the portrait I was looking for : a large glass float in its original net, a ski, and a ship's lifebelt.

While the origins of the float and ski will never be known, the lifebelt came from a converted Norwegian whaling ship, which was wrecked on Westray in 1988 while transporting a cargo of salmon smolts from Oban to Shetland. The ship's steering gear malfunctioned as it sailed past the island, causing the vessel to turn sharply towards the Westray cliffs. Tearing a hole in her bow, the ship quickly sank, but not before the four-man crew took to one of the life rafts. But in a heavy surge of the sea, the life raft also had a gash torn in it. The men scrambled to the shore, when a Coastguard helicopter was soon on the scene and winched them to safety.

29

GEORGE COSTIE and THE FISH HOOSE, Westray

In the vast, gloomy rooms of Pierowall's Big Store, or the Fish Hoose as it's usually known, George Costie's beach-combing finds gather dust. The ever-increasing number of buoys, creel markers, purse-net floats and fenders, ropes and boxes are mostly unwanted now with the demise of Westray's fishing fleet, and now the creel fishermen are few.

Built in 1883, the store was originally designed for a busy fish-curing business, the salting of cod and ling. But the boom was short-lived and, by the early part of the 20th century the building became redundant. Taken over by a coal merchant to begin with, and later housing a weaving business, the Fish Hoose has for the last few decades been filled with the paraphernalia of the fishing industry.

But today in the Fish Hoose, the collection grows, though demand waned long ago. The beachcombing habit is difficult to break. Now belonging to Heddle Costie, the store's future is being discussed in Westray, with ideas to turn the building into flats, or into workshops and offices.

30

TAM MACPHAIL and WHALEBONES, Stromness

With a view south to the ever-changing light and colour over Hoy Sound and the sheer cliffs of Hoy and the Atlantic Ocean, Tam MacPhail sits in his Don garden with an assortment of whalebones. Collected by his late wife, the Swedish-born photographer and artist Gunnie Moberg, the bones were found at Billia Croo, a tiny bay below the house. Gunnie was deeply fascinated by the Orkney shore, observing and photographing it almost every day. 'The rocks have their own seasons', she said, while talking about some of her photographs at an exhibition.

Gunnie pursued a wide range of artistic talents, which included creating this garden at Don. It was here that Gunnie found that she could attract a pair of ravens to the garden from the Black Craig, the sea cliff behind the house, by calling them, and leaving food for them on a wall. And the chair which Tam is sitting in was made by Gunnie . Made out of old flagstones, it is called the George Mackay Brown chair after their friend, the poet and novelist, who was the first person to sit in it.

And Tam, all the while, faithfully runs Stromness Books and Prints, with his dry and wry humour. His advertisements for the bookshop in the local press have been noteworthy. Below an engraving of a galloping horse, for example, boasts the headline "Scotland's Only Drive-In Bookshop". 'Well, people have driven to the door as there's no pavement on the street, and they've been handed a book through the car window'. Another advert is a photograph, taken by Gunnie, of Alisdair Gray with a book on his head, and has the line "Whatever it is, there's a book on it". 'I thought of calling the business Tamazon Books, but no…, and someone suggested Tam iz one…'

31

ELIZABETH SICHEL and SPERM WHALE TOOTH, Sanday

The remnants of old shipwrecks and bleached whalebones litter the coast of Sanday's remote northern headlands, Whale Point and the Holms of Ire, and the Riv. Stretching into the North Sound, the wide sea between Westray and North Ronaldsay, these dangerous rocks are a rarity on this island of long, white sandy beaches and sheltered bays.

It is in this remote part of Orkney, at Upper Breckan, that Elizabeth Sichel and her husband William established a business, in the early 1980s, making garments from wool shorn from Angora rabbits.

In a small bay near their house, around 2002, Elizabeth was out "bagging the bruck", a beach-cleaning-activity which takes place every May involving about 50 islanders. On picking up what she thought was a piece of plastic, Elizabeth discovered she had found the tooth of a sperm whale.

At one time, this tooth would have been prized by whalers for the practice of scrimshaw, where a drawing, often a sailing ship, a whaling scene, or a wife, would be finely carved in the enamel surface. The picture would become clear with a rubbing of lamp or candle black.

32

OLIVER SCOTT with MALLET and COPPER FLOAT, North Ronaldsay

Crofting for most of his life in Orkney's most northerly island, Oliver Scott gained a B.Sc. in Agriculture at Aberdeen University before returning to North Ronaldsay. But to begin with, in those early years he also taught for just over four years at the island school, where there were around eighteen pupils on the register.

After Oliver left his teaching post, he was able to devote more time to the croft, and to family life with his wife Winnie, and to maintaining the island's extraordinary flock of sheep which live entirely on the shore (apart from the ewes that are brought in-by for lambing in the spring), surviving on a diet of seaweed. These sheep, small and tough, and valued for their meat and wool, are restricted to the island shore by a dry-stone wall, thirteen miles in length, and, in places, ten feet in height.

Throughout the year, a number of North Ronaldsay islanders share the labour of maintaining the flock under the authority of a sheep court, to carry out the punding, the communal gathering of the animals for shearing, medicating and selection for slaughter.

Photographed here at the age of seventy-six, in July 2011, Oliver has a huge mallet, found on the shore by his son-in-law recently, and a beautifully corroded copper ball, probably used as a fishing-net float, and found by his father during WWI.

Oliver passed away on Thursday, 31st May, 2012, after a long illness.

33

GLEN FARG LIFEBOAT, North Ronaldsay

Lying at rest now in a sea of tall grass and nettles, a small, clinker-built boat slowly falls apart at Greenwall, in North Ronaldsay. She was originally a lifeboat on the *ss Glen Farg*, an 876-ton cargo ship which was torpedoed on 4th October, 1939, while on its way from Norway to Scotland's east coast ports of Methil and Grangemouth. Shortly after the sinking, the lifeboat was washed up and brought ashore near her final resting place.

Carrying a cargo of general pulp, and chemicals including carbide and ferrochrome, one crew member of the *Glen Farg* was lost in the attack, while the remaining sixteen of the crew were picked up by a destroyer, HMS *Firedrake*. The destroyer herself was torpedoed three years later in the North Atlantic when escorting a convoy to the United States, when one hundred and seventy lost their lives and twenty-six survived.

The ss *Glen Farg* lifeboat was used for a few years after Second World War by one of the North Ronaldsay islanders for creel fishing, but it was soon abandoned and has lain at Greenwall ever since.

34

STEWART SWANNEY and *HANSI WHISTLE*, North Ronaldsay

A farmer all his life on North Ronaldsay, raising cattle and growing grass for silage, and raising three children with Mary, his wife, Stewart Swanney showed me around his Kirbist farm, pointing out a variety of items which came from a wreck, ss *Hansi*, which foundered off shore on 7th November, 1939.

Collected by Stewart's father, the whistle in this photograph from the *Hansi* must be one of the most unusual, if not useful, trophies from a shipwreck. 'It would still work', said Stewart, 'my father blew some air through it using an old Hoover many years ago'. But the ss *Hansi* did provide some objects which have proved useful around the farm to this day, including a forge, kept in a corner of one of the byres, and lighting around the yard.

Typical of island life, Stewart has other occupations on the island. He attends the twice-weekly ferry arrivals and departures at the pier, for example, and before the lighthouse was automated (the tallest land-based lighthouse in Britain) he did occasional light-keeping.

U.S.N.A. LIFEBELT, N Ronaldsay.

35

JIM TOWRIE and DRIFTWOOD BOWL, Sanday

On Jim Towrie's sixty-fifth birthday, his four daughters gave him a wood-turning lathe for his workshop. The gift was out of the blue – he'd never thought about wood-turning before. But Jim took to it almost immediately.

'I'll sometimes find a good tree-trunk on the beach, and I have to rush home for my chainsaw', said Jim. 'And I've just been given this huge lump of pine by a couple of friends. They're regular beachcombers'.

The bowl which Jim is holding in this portrait has been turned from a piece of this driftwood pine. Riddled with large holes made by sea-worms, '...they just add to the character of the thing', says Jim.

When this portrait was taken, it was twelve years since he was given the lathe, and his enthusiasm hasn't waned.

To stop the wood from drying out too much, and too rapidly, Jim seals many pieces of his driftwood collection with molten wax, which also came from the shore. The wax is from the wreck of the *Johanna Thorden*, a Finnish motor vessel which had been on its way from New York to Sweden when it foundered in the Pentland Firth, on Swona's southern tip, in 1937. From the ship's hold came many hundreds of lumps of this wax, which were soon washed up on many Orkney beaches.

36

TOMMY GARRIOCH and WWI GERMAN DESTROYER *B98* HINGES, Sanday

When the First World War German destroyer *B98* ran aground on Sanday's north-east shore, yet another ship had delivered unwittingly to the islanders an unexpected assortment of useful and often valuable objects.

After operating as a supply and mail boat to the interned German fleet in Scapa Flow, *B98* was seized by the British authorities in 1919, a few months after the Armistice, with the intention of scrapping her. But on the way to the scrapyard, B98 broke free of her tow-rope and eventually grounded in the Bay of Lopness, on Sanday's north-east coast.

Easily accessible, lying close to the shore and with only a slight list, *B98* was soon stripped by the islanders of anything which was removable. And so it was for Tommy's father, who managed to salvage a variety of things, including a couple of large brass hinges. This portrait of Tommy at Stangasetter shows one of those *B98* hinges on his garden gate.

Serious salvage work began on *B98* soon after grounding. All that can be seen now at low tide in the Bay of Lopness is a row of heavily corroded turbines and boilers.

37

WILLIE MOWATT and *GIRALDA* LIFEBOAT, South Ronaldsay

On a snowy hillside, on 30th January, 2010, a group of Royal British Legion members performed a simple and moving wreath-laying ceremony at St. Olaf's Cemetery, just outside Kirkwall. The service was held on the 70th anniversary of the sinking of the ss *Giralda* in the Second World War, with the loss of all twenty-three men on board. Eighteen of them are buried in the cemetery.

Seventy years earlier, two Luftwaffe pilots were looking for North Sea convoys and spotted the *Giralda* just off the north-east corner of South Ronaldsay. Two well-aimed bombs and a few minutes of machine-gunning hit the beautiful 2,000-ton steamship about three miles east of Grimness. Carrying a cargo of coal from Ayr to Kirkwall, she was almost at her destination. Abandoning ship , the crew took to their lifeboat, and happened to be seen by a pilot, Captain Henry Vallance, flying past in a Scottish Airways aircraft. He reported what he saw on landing at Kirkwall Airport, and very soon a crowd gathered at the Grimness shore. Just five months into the war, its brutality was about to be brought home suddenly and dramatically to everyone in Orkney.

The ss *Giralda's* lifeboat overturned about a quarter of a mile from the shore. The horrified onlookers, among them a doctor, a nurse, a policeman and coastguards, were helpless, but some people nevertheless bravely attempted to wade in and reach the bodies being washed towards them in case there were any survivors.

'The crew didn't stand a chance! If they weren't killed by the bombing, or didn't drown, then they were dashed against the worst stretch of rocks in South Ronaldsay', said Willie Mowatt, life-long blacksmith, and a teenager at the time of the tragedy. A photograph taken at the time shows a group of people on the Grimness shore, and in the chaotic surf can be seen the upturned lifeboat of the *Giralda*. It was dragged above the shore, out of the way. So there the lifeboat lay for a while, until someone decided to use it. In this photograph of Willie Mowatt is part of that lifeboat, now upturned again and serving as a peat-store at Cleat, at the opposite end of South Ronaldsay to Grimness.

38

JAMES & SARA SCOBIE in RUSNESS COTTAGE, Sanday

The sea is everywhere in Rusness Cottage. It is a unique Sanday bed-sit, where the kitchen, bedroom and sitting-room combine to make one long room, and is filled with sea-finds from the island's long, white-sand beaches.

Since the late 1960s, this cottage has been a remote holiday retreat from London and James Scobie's busy medical practice. And with two children in those early years, the island was for the family a place of peace and relaxation.

For a few years the Scobies were known in Sanday as The Tourists, and were among the first of the many English pioneers who moved to the island to renovate disused or derelict cottages and houses. Many stayed to live in Sanday permanently.

Taking centre stage, on entering Rusness Cottage, is a platform of old fish boxes, found on Lopness Bay, which supports a double bed; to the left, is the kitchen, and hanging from the ceiling near the sink is a long rope of colourful plastic buoys. To the right, the warm, homely sitting-room, where this photograph was taken, is the heart of the home.

Above the mantlepiece and a row of glass floats painted with sailing ships by Willie o'Nevan, hangs an abstract made of driftwood by Sara. And to the left hangs a long wooden board from Sanday's First World War wreck, the German destroyer *B98*. Part of the interned German High Seas Fleet, *B98* was one of the ships scuttled in Scapa Flow by their crews on 21st June, 1919. Having been re-floated and on its way for recycling, the *B98* was on tow when the line broke, and she drifted ashore at Lopness Bay.

39

CYRIL ANNAL and DINGHY, South Ronaldsay

Co-owner of Swona with his brother Martin, Cyril Annal was attendant keepper of the island's minor lighthouse for many years. His duties were to check and maintain the apparatus for the Northern Lighthouse Board. These regular visits were always an opportunity to scan the beaches and geos around the island for anything of interest; and then, around 1980, Cyril spotted a beautiful, clinker-built dinghy lying in Middle Taing Geo.

On reporting the find to the Coastguard, Cyril discovered that the boat came off a service ship to the North Sea oil industry. The ship had gone into a massive wave near Stroma, and when it came out of the other side, the dinghy was gone.

Being told to keep the boat, Cyril used it for many years to attend to the Lowther Rock Beacon, another of the Northern Lighthouse Board's minor lights, off the southern tip of South Ronaldsay.

Perched on top of South Ronaldsay's highest point, Ward Hill, the building housing the boat now was originally a Second World War power station serving one of Britain's coastal defence systems, known as the Chain Home Low Radar Station.

40

CYRIL ANNAL and CEMENTMIXER, South Ronaldsay

Turning driftwood into fiddles, whalebones into fence posts, or scrap metal into machinery - it's all part of island life. It's about using the materials you find on your doorstep, and your doorstep is often the shore. And when nothing is normally thrown away, a graveyard of old machinery, useful for spare parts, can accumulate behind many Orkney farmhouses.

At Stennisgarth, on South Ronaldsay, Cyril Annal has one of these machine graveyards. In this photograph, Cyril and his son Alexander, demonstrate that one of his old machines is still in working order. It is a cement mixer, made out of scrap which Cyril found many years ago on the South Ronaldsay shore.

The bucket of the mixer is adapted from a float once used in a Second World War boom defence system. The booms consisted of long strings of these floats, and they enabled vast metal nets to hang underneath them. Stretched across the southern entrances to Scapa Flow, the system was designed to prevent enemy submarines and torpedoes from gaining access to the British fleet.

41

DAVID HUTCHISON and ROOF TIMBERS, Hoy

Rackwick. The name is Norse, meaning "the bay of wreckage". An O.S. map of Hoy shows that the bay is positioned like a scoop, waiting to catch anything in the ocean travelling north in the Gulf Stream. The beach, as George Mackay Brown wrote, is '...half huge round sea-sculptured boulders, half sand... Out in the bay, like guardians, stand two huge cliffs, The Sneuk and The Too.'

It is in this dramatic location that David Hutchison has lived since 1980, in a house he built himself from the ruins of a byre. 'All the timber for this house came off the shore,' he says, 'and it cost me only twelve pounds to make it habitable.' The original byre is around two hundred years old, and was attached to Muckle House, one of many crofts scattered around the bay. It is perched on a small plateau and, as David says, 'It's like being in the balcony of a first class-stage show'.

Admiring the view from this perfect balcony, and sitting on a milk crate from Greenwich Village Dairy of Long Island, New York (another beach find, of course), I asked David about the flagstones for the roof and the floor. 'All from a quarry near the Old Man of Hoy,' he said. 'I've never been to the quarry and by the way I've never seen the Old Man of Hoy either,' he said.

'About half a million tins of Danish butter cookies were washed ashore not long after I moved here', David said. 'They were lost from a ship sailing from Denmark to the Faroes. Now I still can't face another biscuit. I have to rush past the biscuit section when I'm in the supermarket. And a small television set appeared one morning, covered in barnacles, and sitting on top of a large, red boulder...'

42

NAN TRAILL-THOMSON and LABRADORITE ROCKS, Stromness

In a beautiful barque called *Harmony*, specially built for Arctic waters and encounters with icebergs, Captain Henry Linklater of Graemsay sailed for many summers from London to Labrador for the Moravian Mission - the Moravian Brethren's Society for the Furtherence of the Gospel. With his cousin and ship's mate, Hurricane Jake, on board, the *Harmony* always berthed in Stromness for a few days to take on extra crew and supplies before venturing across the Atlantic.

Usually sailing in convoy, for safety, with ships of the Hudson Bay Company, *Harmony* visited the Mission's trading posts along the coast of Labrador to barter with the Inuit, (to fund the enterprise), and to spread the Christian message. Then, before the advent of winter, *Harmony* crossed the Atlantic once more and returned to Stromness.

On the Inner Holm, the tiny tidal island close to and opposite Stromness, Captain Linklater built a house, a perfect location for a master mariner. Beside the house, now the home of Nan Traill-Thomson, there lies a collection of iridescent rocks, labradorite, once ballast in the bilges of the *Harmony*. Originally dumped on the shore by Captain Linklater, they are much valued for their extraordinarily beautiful blue quality, depending on the direction of the light. Pieces of the former ballast decorate many gardens in Stromness, and the artist Sylvia Wishart had a ring made with a piece of the stone.

At the age of 64, the captain died on the Inner Holm on the 21st December, 1896. His death occurred only a few weeks after his ship, *Harmony*, was taken out of commission and broken up. In Warbeth Cemetery, by Hoy Sound, Captain Linklater's grave is, poignantly, covered with a layer of labradorite chips.

43

FRAMED PHOTOGRAPH of LORD KITCHENER, Stromness Museum

The distinctive First World War recruitment poster in which Lord Kitchener, the Minister of War, demands 'Your Country Needs You!' has ensured that his name and face remain familiar to this day.

But two years into the war, on 5th June, 1916, with Kitchener on his way from Scapa Flow to Russia on a morale-boosting mission, his ship, HMS *Hampshire*, struck a German mine off Marwick Head, Orkney, in a Force 9 gale. The *Hampshire* sank almost immediately with the loss of 643 men. There were only 12 survivors, and Kitchener's body was never recovered. Curiously, the naval authorities refused permission for the Stromness lifeboat and local people to help in the rescue. Rumours and conspiracy theories spread throughout the islands and beyond.

Another mystery, but by contrast a very small one, lies in Stromness Museum. In a cabinet on the ground floor hang photographs and relics related to famous ships active in Orkney during the two world wars. One section, for example, describes HMS *Vanguard*, which exploded after an accident within the ship with the loss of 804* men, while another tells of HMS *Royal Oak* which suffered a torpedo attack with the loss of 833* men. But the mystery is in an old photograph in the section devoted to HMS *Hampshire*.

The picture is a black-and-white portrait photograph of Lord Kitchener in a plain, sea-washed wooden frame. Found floating in the sea just off Marwick Head by a fisherman in the 1960s, the picture was, strangely, only four miles from the position where HMS *Hampshire* went down about five decades earlier.

*Numbers killed have been recently revised : *Royal Oak* 834, and *Vanguard* 843.

His memorial plaq[ue]
in the **HMS HAMPSHIRE display.**
Presented by Mrs Heather Phillips
(nee Potter), Camberley, Surrey.

Lord Kitchener.

44

CARIBBEAN BEANS, Stromness Museum

In an extraordinary book called *A Description of the Isles of Orkney*, printed in 1693, a wide range of topics fall under the scrutiny of its author, a Kirkwall minister called James Wallace.

One chapter begins: 'The people here are generally Civil, fagacious, Circumfpect, and Piously Inclined ... The people are generally personable and Comelie ... The women are lovely and of a beautiful countenance, and are very broodie and apt for generation'.

This rare copy of the book, in Stromness Museum, is open to reveal descriptions of sea beans, which are found occasionally on the Orkney shore, having been carried in the Gulf Stream across the North Atlantic Ocean from the Caribbean.

For centuries, people have treasured these objects and used them as amulets. There is a variety known as "Mary's Bean" or the "Crucification Bean", as it has a cross on one side. The finder would carry one of these beans in a pocket for protection from evil spirits. The bean was also believed to be a safeguard against drowning, as it had survived the sea itself; and a woman would suffer no pain during childbirth if the bean was clenched in the hand.

Pectunculus vul........ ......tundus,
circiter 26........ ....ibus
donatus.
Tellina int......
bitu fe.......
more of.......
lours, and......
riegated, bu......the
same score wit........are
so:
Concha lævis, altera .... parte clauſi-
lis, apophyſi admo........ ......inente, la-
taq; prædita.
Solen ſive conc........ ....; ab
utraq; part........ ......The
Spout Fiſh........
Muſculus ex........ ......m-
mon Muſcl......

On a Log of ....been
some time in the Se........ ......wards
thrown upon the ſhore........ Storm,
I have ſeen thouſands of the *Balani Ron-*
deletij, or the *Concha ....* ; and on
the Rocks every wh........

Balanus cinereus, ...laminis
ſtriatis compoſitus, ...e, altera
teſta biſida rhomboide .... There

WHALE SKULL on LIDDLE'S PIER, Stromness

Washed up on the Yesnaby shore, and then a little later washed down to Billia Croo, on Hoy Sound, this is the skull of a minke whale whose body measured seven metres in length. Found in 2002 by Gareth Davies, the Managing Director of Aquatera, the Stromness environmental consultancy firm, the skull was removed from the whale's body and has stood on Liddle's Pier in the town ever since. Catching the eye of tourists as they explore the streets, the skull is one of many whale-bones which can be seen around the town.

Historically, whalebones in Orkney have been considered very valuable objects in a place where wood is so rare. Huge pieces of whalebones were often used as strainer posts in fencing, and can still be found on some islands today. And probably from the right whale, the largest vertebrae were used as milking stools in many an island cow byre. But Orkney's most beautiful piece of whalebone is the famous Scar dragon plaque. It is a Viking work of art, a carving of two dragon heads, facing each other, and was found in a burial boat in Sanday by a farmer in 1995, after the sea had eroded part of the coast. The plaque is on display in the Orkney Museum, opposite the great, red St. Magnus Cathedral in Kirkwall.

Malcolm Macrae is photographed at a whalebone arch on his Binscarth Farm. 'It was probably put here by my great-great-grandfather Robert Scarth. He drained all these fields and built all the dykes.'

46

NEIL LEASK and SHIP'S KEEL ROOF TIMBER, Kirbuster

Having survived unchanged for at least four centuries, this house at Kirbuster, Birsay (now Kirbuster Farm Museum), with its central hearth and a hole in the roof designed for drawing out the peat smoke, is unique in northern Europe. The last people to live here, until 1961, were members of the Hay family. They would say to any visitors to Kirbuster that the huge beam above the fireplace was 'the keel from Noah's Ark', and that this piece of wood was responsible for keeping the house stable and sturdy for so long.

Neil Leask is one of the museum's custodians, and takes a keen interest in Orkney's history. The great beam across the fireplace 'could have come from a wreck on the Birsay shore', he said, 'and any timber on the Orkney shore was extremely valuable'. A ship's keel, or part of one, the beam stretches across the fourteen-feet-wide room. 'And you'd find it difficult to hammer a nail into it', said Neil. 'And near the house there was a nine-hole putting green, in the 1920s', he said, 'and friends would visit the Hays, and gather here and listen to gramophone records on a Saturday night.'

The room photographed here, with Neil Leask, has probably had a fire burning continuously for over four centuries until the house was abandoned, Neil told me. 'It was considered bad luck to let the fire go out', he said. Peat is cut from the Birsay hill even today for the Kirbuster fire.

47

SUNKEN COAL BARGE, Stronsay

After serving as a coal supply vessel for Stronsay's fishing fleet for about three years, this barge dragged its anchors in Papa Sound during a storm in February, 1933. The boat went aground near Whitehall towards the end of the island's fishing bonanza. At its height, the fishing fleet was made up of about three hundred steam trawlers, and up to four thousand people, mostly women, were employed every season gutting and packing the herring.

This barge was operated in Stronsay by the Peterhead Coal Company. It has an official registration - *A.C.W.11* - and was built in London in 1918 at the instruction of the War Office. The hull was made of concrete in order not to use precious steel at a time when metals were in great demand for munitions. For many years after the barge was grounded, Stronsay islanders were able to help themselves to a free supply of coal from its hold, making a welcome change, no doubt, from digging peats.

I followed the shore of the Oyce of Huip to photograph barge *A.C.W.11*. As I held up the camera, a skein of Orkney's ubiquitous, resident geese flew past, completing the picture.

48

DOUGLAS MONTGOMERY and 'SVECIA' FIDDLE, Burray

Commissioned to write a composition for the St. Magnus Festival, Douglas Montgomery plays a dark, haunting tune called 'Svecia', on his viola in an outbuilding destined to become a recording studio at his Burray home.

'When I wrote this tune', Douglas explained, 'I thought for a long time about the *Svecia* and its tragic journey. I thought about the start of the journey in Bengal, and the ship's home country, Sweden, and about how the passengers and crew felt as they were approaching Orkney. And the wreck itself, and the North Ronaldsay islanders who couldn't reach the people on the *Svecia* in their distress. I considered all of this'.

'Then I picked up the fiddle', continued Douglas, 'and tried to translate all these feelings into the music. The fiddle tone, the timbre, is deep and dark, and the tune is dark, appropriately'. And after Douglas had written the tune, he discovered that he has North Ronaldsay ancestors, the Swanneys.

The viola was made by Colin Tulloch, the Kirkwall instrument-maker, out of alpine spruce from Italy and maple from Bosnia. He, also, has North Ronaldsay blood in him, and is well-versed in the *Svecia* tragedy. Adding the most subtle of touches to the instrument, the end-pin is made of the dyewood from the *Svecia's* cargo, and the four tuning pegs are decorated with it. And the varnish has a pigment to give the viola the colour of the dyewood. 'When you hear that viola', said Colin, almost in a whisper, 'it sounds like it has the spirit of the *Svecia* in it'.

49

NEIL RENDALL and SHIP'S BOW DECORATION, Papa Westray

'It would have come from the shore, from a wreck, there's no doubt', said Neil Rendall, cattle farmer in Papa Westray, (or Papay as the island is known), while he was being photographed for this book. 'It must have lain in the farm bothy for generations until Jocelyn, my wife, found it a few years ago', he said, referring to the carved board in his hands.

This remarkable piece of wood was probably mounted immediately behind a ship's bowsprit, and, if there was one, a figurehead. Fine ornamentation generally reflected the sense of importance or wealth of the ship-owners. And, as sailors were a superstitious breed, they preferred to sail with a vessel which had a fine figurehead, so that good luck would be bestowed on them.

This carving could be the only remnant of a grand ship, a barque or a clipper, which once sailed the wild oceans of the world. Or there could be many beams and spars from the same ship supporting roofs in houses and byres all over Papay.

50

ENGRAVING IN WOOD found by author KEITH ALLARDYCE

While beach-combing along the Bay of Skaill in 1990, I found this piece of wood, a laminated board measuring about 10' x 12'. The drawing, an engraving, is cut through a very thin, black sealant, and is a romantic scene – a woman on the shore looking not at the sailing ship on the horizon, but at a ring on her finger. There's a Scandinavian look about the woman's dress.

_____

In early December, 1978, I threw three message-bottles into the sea from Suleskerry, a tiny island which lies in the Atlantic about thirty-five miles west of Stromness. A remote island, Suleskerry had the distinction of having Britain's (possibly Europe's) remotest manned lighthouse at the time. We had about fifteen acres around us, which we shared with thousands of seabirds, including puffins and storm petrels, and a large colony of seals.

I was local assistant light-keeper at Suleskerry Lighthouse at the time, and was there with my companions, principal keeper John Kermode and first assistant keeper, Alistair MacDonald. Being just before the festive season, each message read "Merry Christmas and a Happy New Year from the three keepers of Suleskerry Lighthouse, December 1978". A few months later, a letter arrived in response to one of these message-bottles, from Mr Groat of Stronsay Post Office. He wrote that his wife had found the bottle 'in the Bay of Housby, opposite the island of Auskerry', and rounded off his reply with 'Safe landings on your western paradise'.

The second response to my three message-bottles came as a surprise in a different way. At the end of February,1979, I was on duty again at Suleskerry when I turned on the radio just after lunch one day. As I poured a cup of tea at the table with John and Alistair, we heard the last few minutes of Radio Orkney's mid-day broadcast. Then the presenter, Howie Firth, announced: 'And finally, I have news of a message in a bottle being found by a lady on the Sutherland coast. It is a message of season's greetings to the finder from the keepers of Suleskerry Lighthouse'. Howie later ended his programme with: 'Well, if you're listening out there lads, don't forget - next time, post early for Christmas!'. .

END WORD

This conclusion to *End Word* is being written in my house on the Northumberland coast, with a view over a small skerry (called a carr, here), and beyond, the grey North Sea. The sittingroom is filled with seafinds, including two huge whale-bone vertebrae which sit on top of a book-case; there are the sun-bleached skulls of a host of seabirds on shelves and a sideboard; and a length of sea-sculpted oak from a trawler which was wrecked in the 1950s less than half a mile away, lies on a window-sill. All these objects have a story behind them, real or imagined.

In his book, *Ring of Bright Water*, Gavin Maxwell wrote movingly about his life at Sandaig, a remote bay on the west coast of Scotland, during the middle of the last century. His accounts of beachcombing describe both the practical and the almost spiritual nature of the activity. For example, after two or three years at Sandaig, having accumulated a wide range of useful objects from the shore, he believed all he lacked for his house was a clothes basket. And within a few weeks '...a clothes basket came up on the beach, a large stately clothes basket, completely undamaged'. On the other hand, some of the jetsam found by Maxwell in those early days at Sandaig reflected much pathos, such as the fire-blackened transom of a small boat, and broken and wave-battered children's toys.

Some finds were so enigmatic that around them he could only 'weave idle tapestries of mystery'.

But it wasn't only the man-made objects discovered along the coast which Maxwell found so inspiring. His imagination could be fired on finding the remains of a bird or an animal amongst the sand and wrack. From these remains, he wrote, '...rise images the brighter for being unconfined by the physical eye. From some feathered mummy, stained and thin, soars the spinning lapwing in the white March morning ... and the broken antlers of a stag re-form and move again high in the bare, stony corries and the October moonlight'.

Beachcombing, this honourable tradition, is about using the imagination, and it's about curiosity and surprise. Beach-combing might even stem from a primeval need to be near the sea, to hear and smell the sea, and to feel and touch its immensity and enduring mystery.

Keith Allardyce.

On the Northumberland coast, August 2012.